Mehdi BOUDOUH
Brahim Elkhalil Hachi

Towards a New Highly Efficient Handling Technology

Mehdi BOUDOUH
Brahim Elkhalil Hachi

Towards a New Highly Efficient Handling Technology

Efficiency-Performance-Safety

ScienciaScripts

Cover image: www.ingimage.com

This book is a translation from the original published under ISBN 978-620-6-71433-0.

Publisher:
Sciencia Scripts
is a trademark of
Dodo Books Indian Ocean Ltd. and OmniScriptum S.R.L publishing group

120 High Road, East Finchley, London, N2 9ED, United Kingdom
Str. Armeneasca 28/1, office 1, Chisinau MD-2012, Republic of Moldova, Europe
Printed at: see last page
ISBN: 978-620-7-75638-4

SUMMARY

Forklift trucks are frequently used for handling, storage, stacking and unstacking operations. To this end, the industrial sector using this type of machine requires a number of conditions, including safety, performance (power and capacity), reliability, availability, comfort, price and aesthetics. In this work, we have tried to propose and verify certain techniques with the aim of researching and selecting the standards applicable to forklift trucks to ensure that they are in good condition, that they are used correctly, and to identify the risks associated with the use of these machines.

CONTENTS

SUMMARY 1

GENERAL INTRODUCTION.. 3

CHAPTER I 4

CHAPTER II.. 31

CONCLUSION 46

BIBIOGRAPHICAL.. 47

GENERAL INTRODUCTION

Self-propelled forklift trucks - so many different types of stacking and handling equipment - have become a regular feature of the industrial and commercial landscape. There are few places where they don't make their contribution. Their great adaptability and speed of implementation make them a strategic investment, enabling companies to 'stick' to their markets and keep pace with changes.Unlike fixed installations, a "trolley" solution means that investment costs can be spread over time, and can be easily adapted to numerous changes in the manufacturing process, production machines, product characteristics, packaging and working hours.Leasing is particularly well suited to this type of investment. It makes it possible to reserve finance for other essential investments. It also makes it possible to cope with peaks in activity and to compensate for breakdowns and other unforeseen events [1]. In this work, we are motivated by an objective set by the GERMAN company, which specialises in the manufacture of handling and stacking equipment (the only one of its kind in Algeria). This objective consists of developing and improving the cab of the 7-tonne machine to give it greater comfort, aesthetics, safety and ergonomics, while respecting the quality/price ratio.This dissertation is divided into three chapters. Chapter one gives definitions of handling and stacking, whether manual or mechanised. The second chapter is devoted to shedding light on a specific aspect of handling and stacking.forklift truck. Finally, chapter three presents a technological study aimed at improving the driver's cab in terms of aesthetics, safety, ergonomics and value for money. This work concludes with a general conclusion.

CHAPTER I
HANDLING AND STACKING

I.1. Introduction

Handling refers to the transport or support of a load that requires physical effort by one or more people. This effort may be made to lift, place, push, pull, carry or move the load. Because of the conditions in which it is carried out, handling may involve risks to workers' health and safety.In some logistics management chains, particularly in labour-intensive production sectors, materials handling is of vital importance. It enables heavy goods to be lifted or moved. In this case, the use of forklift trucks or other equipment designed for this purpose is essential.In small and medium-sized companies, a single person is usually responsible for handling management. For large companies, a special department is often set up: this is the plant handling department.This group of people is responsible for improving the company's performance by taking direct action on the layout of machines, the organisation of workstations and the optimisation of handling times [2]. The aim of all companies is to provide the necessary means and equipment to enable handling to be carried out in complete safety.

Today, handling refers to any operation: [3]

- transporting or supporting a load, including lifting,

- installation,

- thrust,

- traction,

- wearing or moving

I.2. definition of handling

Handling is the movement of goods, industrial products or loads over a short distance, see Figure I.1 [4].

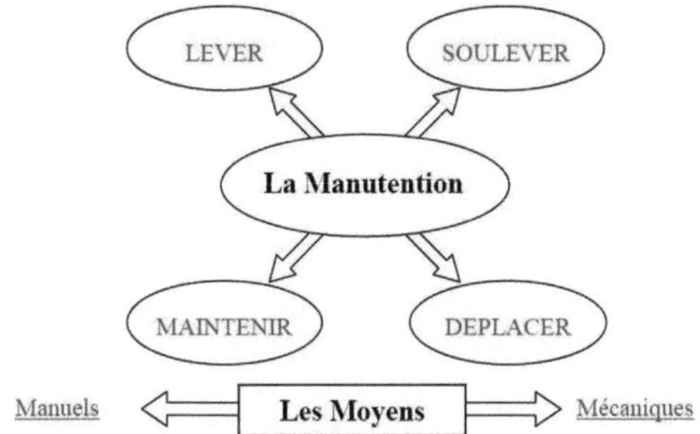

Figure I.1: presentation of handling and their means

I.3. types of handling

There are two necessary types of handling: manual and mechanical. [5]

I.3.1. Manual handling

I.3.1.1. Definition

Manual Handling means any operation of transporting or supporting a load, including lifting, placing, pushing, pulling, carrying or moving, which requires the physical effort of one or more workers.

I.3.1.2. Manual handling methods

Manual handling requires the physical effort of one or more people and, because of its characteristics or the conditions in which it is carried out, may involve risks to workers' health and safety.Manual handling affects a large number of companies. Every year, thousands of people are permanently affected by these occupational illnesses or by a work-related accident. Certain sectors are particularly affected, such as construction and public works, logistics, metallurgy, mass retailing, the agri-food sector and transport. [6] When handling loads, the physical effort required of our bodies puts strain on the spine, joints and muscles, and increases cardiac activity. These efforts are not without consequences for the body and can cause specific and painful pathologies, including musculoskeletal disorders (MSDs). These are injuries to the periarticular areas and all the segments of the body. They are most often linked to incorrect movements when moving loads, to ill-adapted workstations and to the performance of repetitive tasks with low amplitudes. [7]

A. THE LEVERS

These are long, rigid bars designed to move, support or lift other bodies.

- **Roller levers**

Often used simultaneously at different points of the load to effect a displacement. See Figure I.2.1.

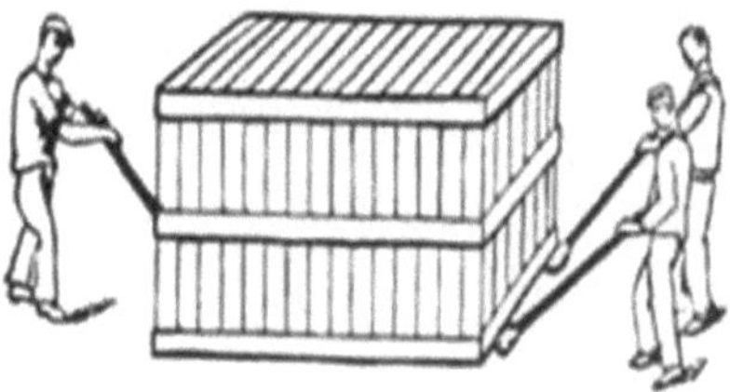

Figure I.2.1: Roller levers

► **Lever handles**

This type of roller lever has a sliding steel support which rests on the ground, holding the lever in its horizontal position for lifting the load, see Figure I.2.2.

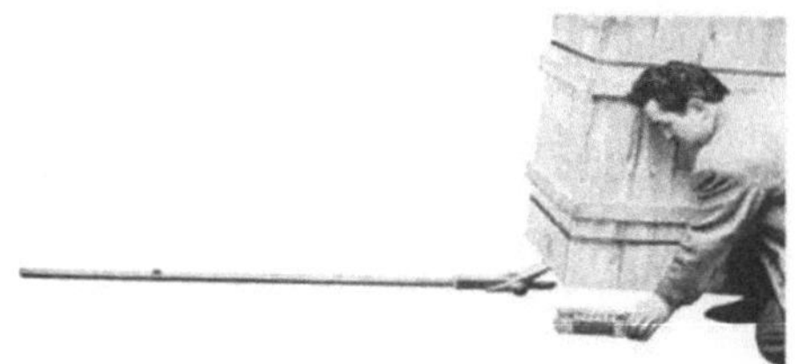

Figure I.2.2: Levers with crutch

The technical specifications are :
Lifting height ;low height : 125 mm, Travelling distance : over a short distance, Lifting capacity : 500 kg to 1 000 kg

B. THE CRICS

These devices are designed to lift loads by reducing the effort required to manoeuvre them. Depending on the leveraging principle used, these devices are divided into two groups:

❖ **Rack and pinion or gear-driven equipment**

They consist of a rack driven by a pinion. A crank is used to raise the load support rack assembly. A pawl is used to maintain the position. (See Figure I.3.1).

❖ **Hydraulic equipment**

These very powerful jacks can lift very large loads. The operator activates the hydraulic pump to lift the load. (See Figure I.3.2).

Figure I.3.1: Rack and pinion jacks

Figure I.3.2: Hydraulic jacks

The technical specifications are : Lifting height : from 150 mm to 1 metre, Travel distance : over a short distance, Lifting capacity to 20 tonnes (gears)1 to 100 tonnes (hydraulic)

C. LES CHANDELLES

Wedging stanchions are used to support loads securely and are usually made of metal. There are three types (see Figure I.4.a, b, c):

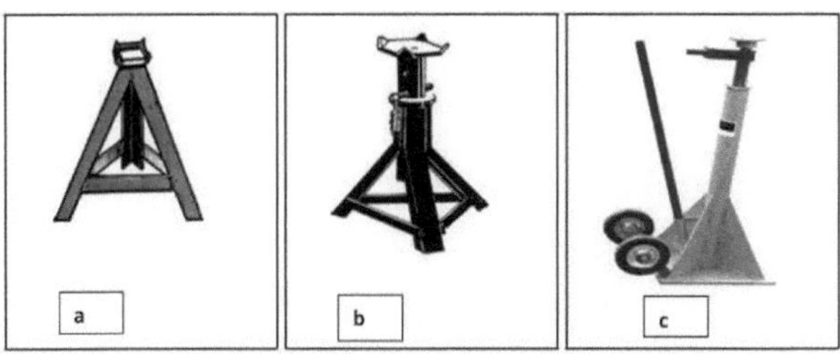

FigureI.4:Stands,a) :Fixed height,b) :Adjustable height,c):Adjustable height

The technical specifications are :

Solid support; height: 300mm to 2 metres, capacity: 2 to 20 tonnes

D.THE ROLLERS

These very robust machines are used to move large loads by rolling. They consist of a sturdy metal frame resting on rollers, see Figure I.5.

Figure I.5: Handling roller

Note: The rollers must point in the direction in which the load is moving. They can be swivelled to facilitate handling.

Their role is to move medium and heavy loads. Capacity from 500 kg to 80 tonnes.

E.Manual pallet trucks

They can be used to lift and move loads over medium distances, see Figure I.6.

Figure I.6: Manual pallet truck

For use only on level ground or ground with a gradient of less than 2%
Maximum load moved: 600 kg for a single man, 360 kg for a single woman

F. MANUAL HOISTS

They allow loads to be lifted over a wide range but require a means of attachment (lifting rings, slings, etc.), see Figure I.7.

Figure I.7: Chain hoist

Lifting height: maximum height: 3 metres. Capacity: 500 kg to 2 tonnes (chain-driven) , 750 kg to 6 tonnes (lever-driven)

G. WORKSHOP CRANES

They allow loads to be lifted and moved, but require a means of attachment, as with hoists (see Figure I.8).

Figure I.8: A manual workshop crane

They consist of a telescopic boom and a hydraulic cylinder. The hydraulic cylinder is powered when the pump is operated by the operator. Lifting height: 1.5 metres, Moving loads over short distances, just a few metres. Capacity: from 500 kg to 2000 kg.

H. OVERHEAD TRAVELLING CRANES

These are lifting devices designed to lift and move loads; they move on parallel tracks, and their gripping device (hook or other lifting accessory) is suspended from a lifting mechanism by means of a cable and pulleys. (winch or hoist) likely to move perpendicularly to the device's raceways, see figure I.9.

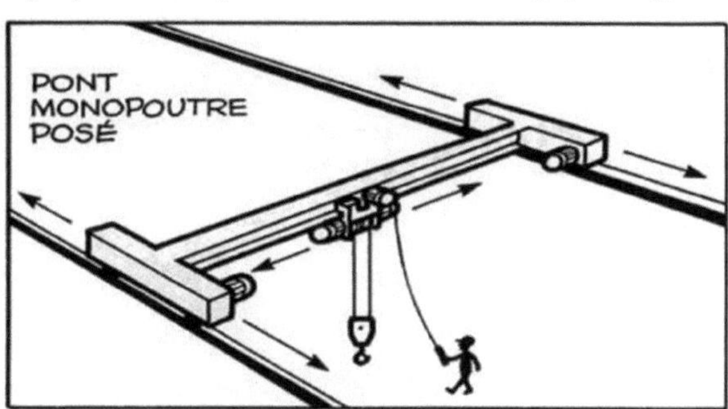

Figure I.9: Single-girder bridge installed

Lifting over several metres ;

Displacement according to the length of the tracks; Lifting capacity is from : 1 tonne to 20 tonnes or more.

I. THE SLINGS

Some lifting devices require a connecting device between the load and the hook of the device, see figure I.10.

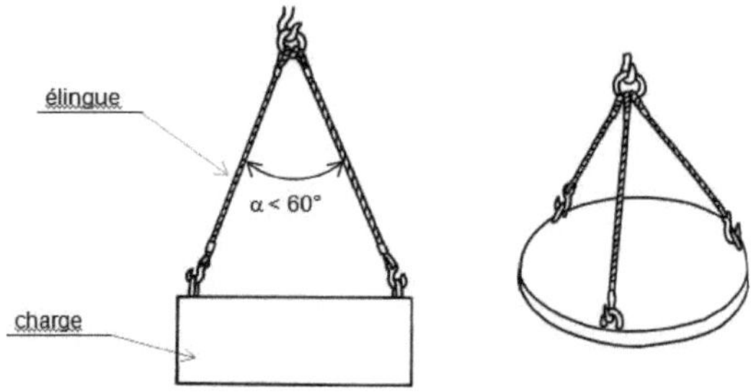

Figure I.10: sling carrying a load

Slings must always be used with equipment that is in perfect condition and, above all, in compliance with all safety rules. Depending on the work to be carried out, slings come in different forms, for example: (See figure: I.10.a, b, c).

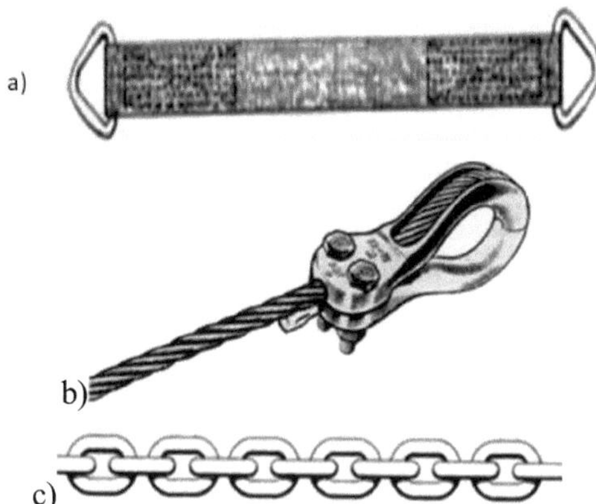

Figure I.10.1 : a) flat woven sling, b) steel cable sling c) chain sling

I.3.2. Mechanical handling

I.3.2.1.Definition

Mechanical handling involves the use of lifting and transport equipment and avoids the risks associated with manual handling. However, it also gives rise to risks associated with the movement of machinery, the load being handled or the handling equipment.

I.3.2.2. Mechanical handling equipment[8]

Handling is necessary when the work requires a high level of attention, hence the use of handling equipment such as forklifts. These can handle and transport any type of load. With its variable size and load capacity, it can move goods without risk.The aim of all companies is to provide the resources and equipment needed for safe handling. Originally, materials handling was the activity of moving parcels and pallets by hand. Thanks to technological advances, it is now possible to carry out handling work using more efficient tools. The introduction and use of handling equipment enables companies to improve productivity. Investing in materials handling equipment makes it possible to be profitable and to beat the competition. These include the following:

I.2.2.2.1.Pedestrian-controlled forklift trucks [9]

A.Pallet trucks, platform trucks

- **Features**

The hydraulic lifting movement is transmitted to the support rollers by a set of levers and connecting rods. Energy is supplied by an electric battery. Some trucks have an integrated charging station. Length of fork arms and overall width of fork arms to be determined according to the dimensions of the pallets to be transported. In the case of platform lifts, the fork arms are replaced by a platform.

• **Performance**

Fork lift height: 300 mm. Travel speed: 6 km/h maximum. Gradeability: varies from 20% unladen to 5-10% laden. Most trucks have a capacity of 2,000 kg, and some are even designed for 3,000 kg.

• **Common uses**

Not very intensive use.
For loading/unloading lorries.
For transporting palletised loads over short distances (30 m) in factories, shops and warehouses.

• **Benefits**

Simple equipment that does not require a driving licence, except for whether a folding platform is suitable.
Compact equipment for use in confined spaces.
Low acquisition cost.
Their low weight means they can be used on floors of low strength, such as storeys and lorry beds.

• **Disadvantages**

Use limited to short distances because the operator is on foot. Exposes the operator to risks:

- of collisions with ride-on forklift-trucks, since it is required to walk down the same aisles,

- entrapment or crushing of the body or a limb against an object obstacle by the chassis, drawbar or wheels. The ground must be in good condition, flat and free of holes.
The use of a pallet truck with platform folding platform is not recommended because :

- the risk of the driver being ejected when cornering and braking,
- lack of protection for the driver,
- the poor ergonomics of the driving position.

This hybrid equipment is dangerous to use routinely (see Figure I.11) and requires a driver's licence.

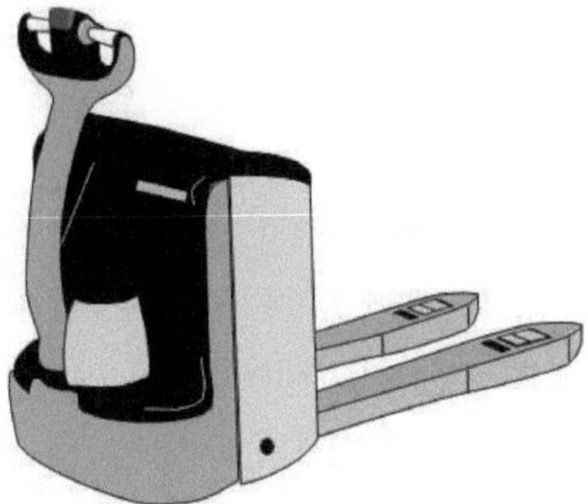

Figure I.11: Pedestrian-controlled mechanical pallet truck

B. Stackers

• Features

Devices derived from pallet trucks, equipped with a lifting assembly for lift the load.

Consider :

- Overlapping forklift trucks, which make up the bulk of the fleet,
- trolleys with framing arms,
- overhanging stackers.

Energy is supplied by an electric battery. Some trucks have an integrated charging station.

• Performance

Travel speed: 6 km/h maximum. Lift speed: varies from 0.25 m/s unladen to 0.10 m/s laden.

Permissible ramp: varies from 10% unladen to 5 to 10% laden. Truck capacities range from 1,000 to 1,600 kg, with a centre of gravity of 600 mm. Some are even designed for 3,000 kg. Lift heights can reach 5 m.

• Common uses

Not very intensive use. For stacking palletised loads to be transported over short distances. For lifting heights of up to 3 m.

• Benefits

Simple equipment that does not require a driving licence, except for whether a folding platform is suitable. Their very low dead weight means they can be used on floors of low strength. Their small footprint and manoeuvrability mean they can be used in confined spaces. Relatively low cost.

Framing arm trolleys are very stable.

• Disadvantages

The capacity of these trucks decreases rapidly with lift height. From a lift height of around 3 m this type of truck is sensitive to sideways tilting. Overlapping fork stackers require pallets to be picked up in one direction only. Use limited to short distances, as the operator is on foot.Exposes the operator to risks:

- of collisions with ride-on forklift trucks, since he is required to in the same aisles,

- entrapment or crushing of the body or a limb against an obstacle by the chassis, drawbar or wheels,

- of falling objects handled at height, as forklift-trucks do not have driver protection.

The floor must be in good condition, flat and free of holes (see figure I.12).

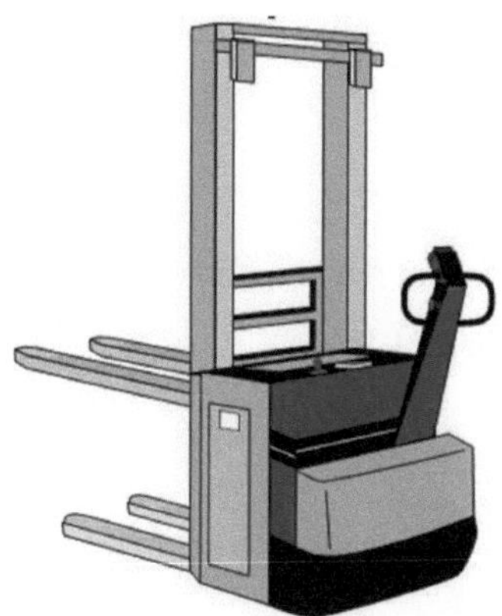

Figure I.12: Pedestrian stackers

C. Tractor

• **Features**

Forklifts developed from electric-powered pallet trucks. replaced by a trailer coupling system.

• **Performance**

Travel speed: 6 km/h.

Towed load capacity: 2,000 kg.

• **Common uses**

Moving small loads over short distances.

• **Benefits**

Extremely manoeuvrable, they can be used in conjunction with other equipment to clear workstations or move through congested areas.

• **Disadvantages**

The floor must be in good condition, flat and free of holes.

Exposes the operator to risks:

- collisions with forklift trucks

Worn, since it is likely to be used in the same aisles,

- entrapment or crushing of the body or a limb against an obstacle by the chassis, drawbar or wheels.

• **Benefits**

Simple equipment that does not require a driving licence, except for if a folding platform is suitable (see figure I.13).

Figure I.13: Pedestrian tractor

I.2.2.2.2. Pedestrian-controlled forklifts [10]

The use of any ride-on forklift truck requires the driver to hold a driving licence issued by the company manager.

A. Trolleys

• **Features**

The chassis of these trucks is designed with a platform on which the load is placed. These trucks are mainly powered by electricity.

• **Performance**

Travel speed: 15 to 25 km/h. Load capacity: up to 2,000 kg.

• **Common uses**

Servicing large workshops, inter-building links, hospitals, port facilities, road services in major cities.

• **Benefits**

Equipment well adapted to needs.

• **Disadvantages**

Very specific equipment.

Requires another means of handling for loading.

(See figure I.14)

Figure I.14: Load-bearing trolleys

B. Tractor trolleys

• **Features**

Tractors have a towing hook, automatic or not, to pull a train of trailers.

These trucks can be thermal or electric.

The tractor braking system is designed to stop the trailer train. It is generally possible to fit a trailer brake control device to tractors.

• **Performance**

❖ Tractors derived from pallet trucks :

- Travel speed: 5 to 12 km/h.
- Towing capacity from 1,500 to 3,000 kg.

❖ Specific tractors :

- Travel speed: 15 to 25 km/h.

- Towing capacity from 6,000 to 20,000 kg (excluding specific airport tractors designed, for example, to pull aircraft).

• **Common uses**

In certain companies, railway stations and airports, to pull a train of trailers.

• **Benefits**

Equipment well adapted to needs.

• **Disadvantages**

Very specific equipment(See figure I.15 , I.16)

Figure I.15: Tractors derived from pallet trucks

Figure I. 16: Specific tractors

C. Pallet trucks

• **Features**

Trucks derived from the pedestrian pallet truck, but where the operator is carried. They require a driving licence. Depending on the truck, the operator may be standing or seated. There are the following types:

Stand-on operator with drawbar,

-Driver standing on steering wheel,

-Rider seated on a steering wheel. Energy is supplied by an electric battery.

The length of the fork arms and the overall width of the fork arms must be determined according to the dimensions of the pallets to be transported.

- **Performance**

Travel speed: 8 to 12 km/h. Gradeability: varies from 20% unladen to 5-10% laden.Most trucks have a capacity of 2,000 kg, and some are even designed for 3,000 kg.

Fork lift height: 300 mm.

- **Common uses -Intensive uses.**

For loading/unloading trucks and wagons. For transporting palletised loads in factories and warehouses over distances of more than 50 m.

- **Benefits**

Compact, manoeuvrable trolleys for use in confined spaces. Cost low acquisition cost.

- **Disadvantages**

Stand-on, sit-on trucks with the steering wheel and the driver's station at right angles to the truck often have a compact driver's station, which is sometimes not ergonomic enough. The compactness of the driving position exposes the driver, who often has part of his body protruding beyond the truck's frame.

There is a high risk of accidents when using these machines.

The floor must be in good condition, flat and free of holes (see figure I.17).

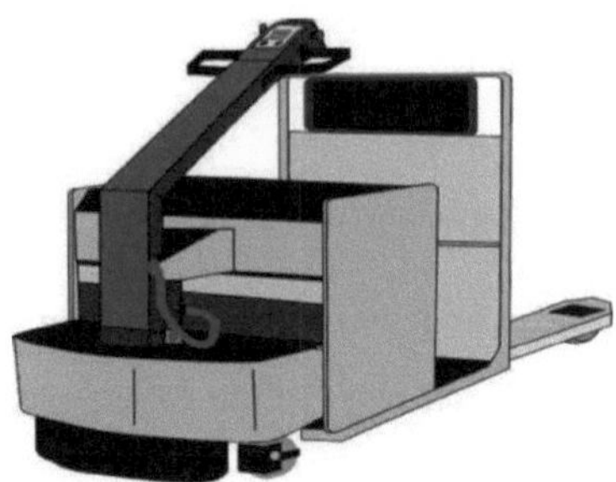

Figure I.17: Ride-on pallet trucks

D.Stackers

• **Features**

Equipment with a driver station that is usually upright, equipped with a lifting column to lift the load. These trucks do not work overhanging.

Consider :

-Overlapping forklift trucks,

-trolleys with framing arms.

Energy is supplied by an electric battery.

• **Performance**

Travel speed: 8 to 12 km/h.

Lift speed: varies from 0.25 m/s unladen to 0.20 m/s laden.

Gradeability: 10-20% when unladen, 5-10% when laden.

Trolley capacities range from 1,000 to 2,000 kg, with a centre of gravity of 600 mm.

Lift heights can reach 5 m.

• **Common uses-Low intensity use.**

For stacking palletised loads in factories and warehouses in confined spaces.

• **Benefits**

Compact, manoeuvrable trolleys for use in confined spaces.

• Disadvantages

The compactness of the driving position often results in poor ergonomics and exposes the driver, who often has part of his body protruding beyond the truck's frame.The floor must be in good condition, flat and free of holes.
Stackers with overlapping forks have to pick up pallets in one direction only and require a clear floor to stack in a rack (see figure I.18).

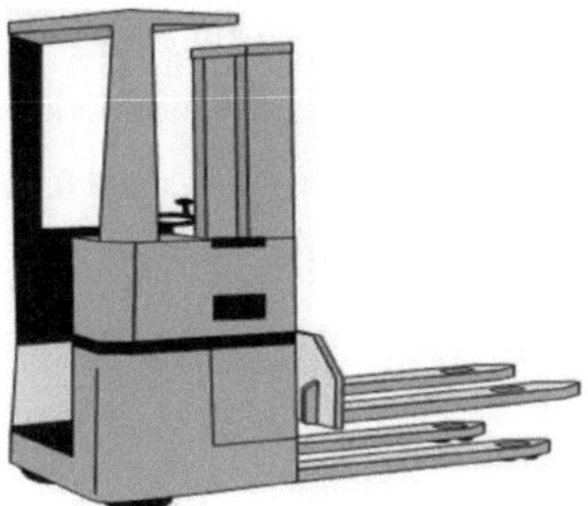

Figure I.18: Pedestrian stackers

E. Cantilevered forklift trucks

1. Front plug

• Features

The forklift unit and load are cantilevered in relation to the front axle, with the counterweight at the rear of the truck providing stability. The operator sits in the direction of travel.

A distinction is made between :

❖ Electric and thermal forklift trucks.

- Electric trucks are powered by a battery.
- Thermal forklifts are powered by a fuel: diesel, petrol or LPG.

❖ Rear turret trucks and four-wheel trucks.

-Rear turret trucks can have either the front drive axle or the rear drive axle. the rear power steering turret.

Four-wheel axle trucks have a front drive axle and a rear drive axle.

Director.

• **Performance**

Travel speeds are in the region of 15 to 20 km/h for forklift trucks. electric forklifts and 20 to 25 km/h for internal combustion forklifts.

Elevation speeds are generally between 0.25 m/s and 0.40 m/s. The capacity of these trucks ranges from less than 1,000 kg to 50,000 kg.

The most common lift heights vary from 3 to 6 m and can reach 10 to 12 m.

• **Common uses**

For moving, transporting and lifting loads in all sectors unless the ground requires an all-terrain vehicle.

The trucks with the highest capacities are used in heavy industry or at ports to move containers using spreaders.

• **Benefits**

Robust equipment, generally designed for intensive, multi-purpose use.

Facilitate handling within the company. Used whenever there is a break in the flow.

• **Disadvantages**

The risk of accident when using these machines is high. In some cases, the load can obscure forward visibility. Release of pollutant gases by thermal forklift trucks not fitted with a device treatment plant. The noise level of thermal forklift-trucks is higher than that of electric forklift-trucks (see figure I.19).

Figure I.19: Front-loading cantilever forklift trucks

2. Side grip

- **Features**

Compact electric trucks with retractable masts positioned between the axles, enabling loads to be picked up and lifted from the side of the truck and placed on the truck's load-bearing platform.

- **Performance**

Capacity from 1,000 to 3,000 kg for stacking long loads up to 8 m high.

- **Common uses**

Trolleys designed to handle long loads (profiles, tubes, boards, etc.).

- **Benefits**

Driving through narrow aisles in workshops and warehouses.

- **Disadvantages**

Can only stack on one side. Requires a level floor in good condition, without hole.

Note: This should not be confused with special side-gripping thermal forklifts designed to pick up heavy, moving loads in the open air on terrain that is often undeveloped. (See figure I.20).

Figure I.20: Side-loading cantilever forklift trucks

F. Warehouse trolleys

All these trucks are generally used in shops on floors that are in good condition, flat and without holes. They are powered by electricity and fitted with tyres.

1. Reach trucks

• Features

The chassis of the truck consists of a driver's station at right angles and two arms in which the lift assembly slides. To move around, the lift assembly is retracted, bringing the load within the support polygon and limiting the overall length of the truck.

• Performance

Travel speed: 12 to 18 km/h. Low permissible ramp. The capacity of these trucks varies between 1200 kg and 2000 kg. Lifting heights of up to 10 m are possible. These machines can be fitted with a device that memorises pallet placement and removal levels.

• Common uses

Trolleys used in warehouses, between distribution depots, in cold rooms, etc., for stacking in metal shelving installations where maximum use of volume is required.

• Benefits

Compact trucks for minimum stacking aisles and optimum use of storage volume. Good stability makes stacking at great heights easier, while ensuring high residual capacity.

Good driving visibility.

• Disadvantages

Driving position across the truck, which requires training time to adapt compared with a traditional trolley. The driver is exposed in the event of an impact due to the layout of the driving position. In addition, depending on his build, part of his body may protrude from the truck's frame (see Figure I.21).

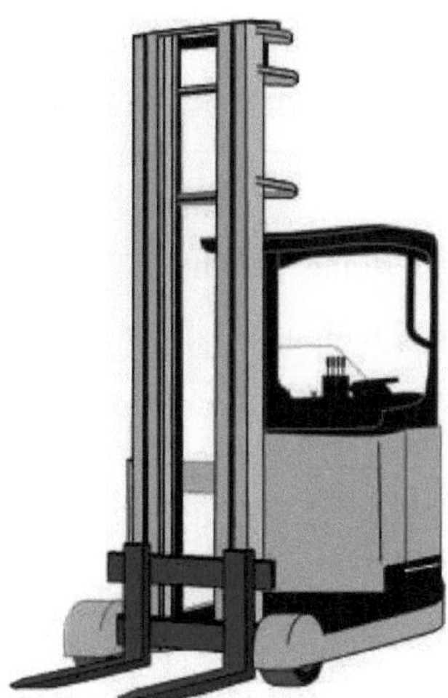

Figure I. 21: Reach trucks

2. Bi-directional and tri-directional trucks

• Features

The bi-directional load pick-up trolleys are designed to be able to pick up the load and place it on one side or the other. Tri-directional load pick-up trucks take the load frontally or laterally from one side or the other.The trolleys are often guided in the racking either by a mechanical roller-rail system or by a wire-guidance system.The trucks are generally equipped with a stacking height selection device and sometimes with a device that automatically positions the forks facing the storage bin.On some forklift trucks, the driving position is raised with the forks.

• Performance

Travel speed: 10 km/h. Lift speed: 0.4 m/s.

Lift heights of 10 m and more.

• Common uses

Trucks used in high-bay metal storage facilities with intensive load rotation.

• Benefits

High productivity. Very high storage heights.
Basic infrastructure compared with installations served by stacker cranes. Lift trucks can be used for order picking.

• Disadvantages

When using these trolleys, the floor must comply with the following specificationsresistance, horizontality, planimetry...Drivers must be trained and familiar with the regulations. to drive safely. Special measures must be taken and staff must be trained to avoid the risks of accidents that are specific to this type of equipment: collisions when leaving aisles, pedestrians colliding in aisles, etc. (see figure I.22 , I.23)

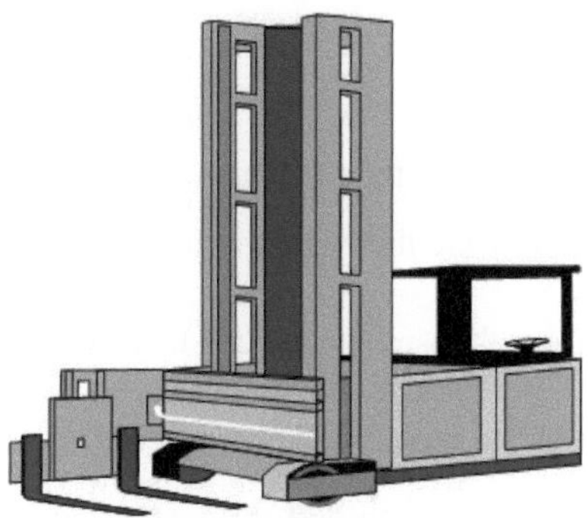

Figure I. 22: Bidirectional trolleys

Figure I. 23: Three-way trolleys

I.4. The consequences of handling :

Even if it is very well designed, handling leads to :

• product damage

• accidents

• a need for storage areas (increase in surface area and taxes) [11]

I.5. The aims of handling :

Handling must enable parts to be moved from one workstation to another on the production line, in order to keep the company running:

• shop raw materials in warehouses - introduce these materials into production

• supply workstations (during the manufacturing process)

• remove the finished product and store it

A handling operation consists of 3 stages:

• aying and securing the part

• travel

• removal of the part to its new location[12]

CHAPTER II

THE FORKLIFT TRUCK

II.1. Introduction

In this section, we try to provide: a definition of a forklift truck, describe the functions and external environment of this machine, research and select the standards applicable to forklift trucks to ensure that they are in good condition and are used correctly, identify the risks associated with using these machines.

II.2. Definition

industrial trucks are wheeled vehicles having at least three wheels with a motorised or non-motorised drive mechanism, with the exception of vehicles running on rails, which are designed to carry, pull, push, lift, stack or arrange in shelves any type of load and which are controlled either by an operator or by a driverless robot [10].The forklift truck is not a piece of transport equipment, but a lifting device designed to pick up loads, transport them over relatively short distances and place them on the designated site.

II.3. Characteristics of a forklift truck

Forklift trucks have characteristics that are often unfamiliar to pedestrians and sometimes underestimated by forklift drivers:[11].The compact shape of the forklift truck does not suggest that it is very heavy - the equivalent of six cars.

Faced with this powerful machine in motion, even at low speed, a pedestrian is out of luck. to escape unharmed in a collision, and even less so if he is trapped. between the carriage and an object.

1- a forklift truck always turns using its rear wheels (axle steering axle) while the front wheels are driven (drive axle).

2- A forklift truck turns more easily loaded than unloaded because of its counterweight, which balances it.

3- a forklift truck operates as much in reverse as in forward motion before.

4- The forklift driver drives with one hand and operates the controls with the other.

5- a forklift truck weighs about the equivalent of six cars (see Figure II.1.)[11].

6- a forklift truck has no suspension.

7- Forklifts, whether loaded or not, have a high centre of gravity.

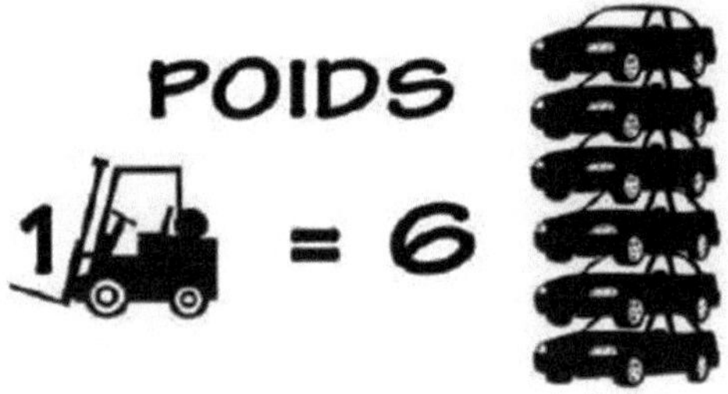

Figure II.1: Weight comparison between forklift truck VS car.

II.4. Technical description of a forklift truck

Forklift trucks are made up of essential components found in all types of forklift.

II.4.1. Technical diagram of a forklift truck

Figure II.2 shows the general components of a forklift truck. [11] :

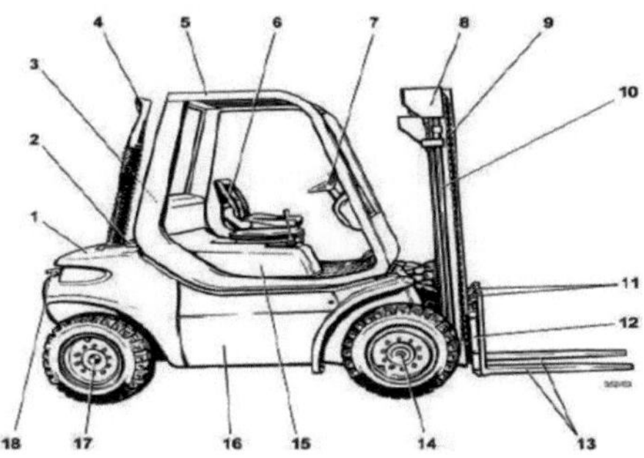

Figure II.2: General components of the forklift truck.

The nomenclature of the trolley shown above is presented in the following table: (table II.1).

number	Component
1	Counterweight
2	Radiator cover
3	Battery box
4	Secondary silencer
5	Protective hoop
6	Driver's seat
7	Steering wheel with dashboard
8	Lifting mast
9	Lift mast chains
10	Lift cylinder
11	Fork arm lock
12	fork carriage
13	Fork
14	Engine bonnet
15	Chassis
16	Steering axle
17	Coupling device

Table II.1 The components of a forklift truck.

II.4.2. Description of forklift components[12]

A. The mast: The mast is the vertical assembly of metal sections that enables the load to be raised, lowered and tilted. The mast can be operated hydraulically: it consists of one or more jacks and interlocking rails that slide into each other via intermediate rollers.

B. The load-carrying carriage: to which the metal fork(s) or optional equipment are attached, moves along the mast by means of chains, or by being attached directly to the hydraulic jack. Generally, theThe load carrier, which is mounted on bearings, is guided and moves between the two rails on the mast.

C. Forks: The forks are the L-shaped arms that hold the load. The rear vertical part of the forks is usually attached to the load-carrying carriage by means of a hook or latch. The horizontal portion of the forks is tapered to facilitate insertion into or under the load, and can be moved horizontally and vertically by means of hydraulic cylinders.

D. A rear load support or backrest: This is fitted when the load is higher than the top of the load carrier; it is a rack-like extension bolted or welded to the load carrier deck to prevent the load from moving backwards.

E. The cab, with space for the driver or operator, contains all the movement actuators: the control pedals, the steering wheel, the hydraulic control buttons and levers to steer the truck, a dashboard with warning lights and a load diagram informing the driver of the weight not to exceed according to the dimensions of the load and the height of lift. The cab can be opened with a simple overhead guard or closed.

F. Diesel propulsion: The diesel forklift engine is identical to a conventional engine fitted to cars for low power ratings, or to trucks and boats for high power ratings. The diesel forklift truck is for outdoor use only. It offers the power of a diesel engine for all load capacities.

G. Transmission: The industrial truck will be propelled using two techniques, either by hydraulic motors (hydrostatic transmission) or by means of a converter (hydraulic coupler), (see figure II.3) [13].

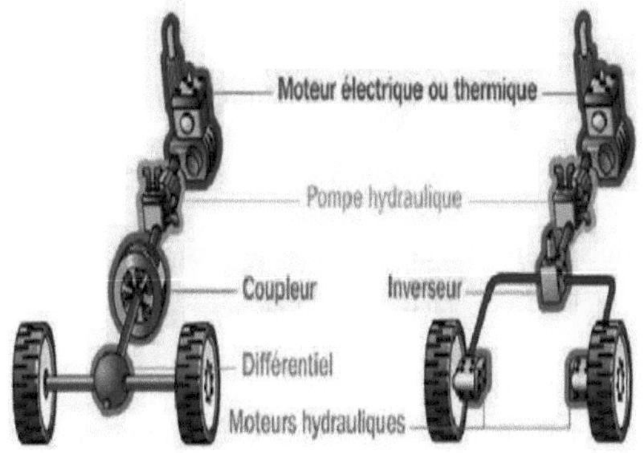

Hydraulic coupler Transmission hydrostatic

Figure II.3: Different types of trolley propulsion.

H. A counterweight: This is a metal mass attached to the chassis at the rear of the truck, needed to compensate for the mass of the load.

I. Wheels with a solid tyre: The tyre of a solid wheel is a casing crimped onto the wheel which completely covers its surface in contact with the ground. The tyre is comparable to a solid tyre, with a low bead height. The tyre has low deformation and absorption capacities, enabling it to overcome unevenness. Solid wheels with solid tyres are suitable for use on prepared ground.

J. The battery: A motorised forklift will require the use of a starter battery. This battery must deliver a high current to activate the starter, but only for a very short time.

II.4.3. How it works

II.4.3.1. Engine

It is driven by a 4-cylinder direct-injection diesel engine which drives the truck's hydraulic pumps at load-dependent rpm. Cooling is by a closed circuit with an expansion tank[13].

II.4.3.2. Hydraulic system (Hydrostatic transmission)

The hydraulic system consists of a variable displacement hydraulic pump and two constant displacement hydraulic motors, mounted to form the compact axle, plus a tandem pump (constant displacement) for the lifting and steering hydraulics. The direction and speed of travel are controlled by two pedals which act on the variable displacement pump. The constant displacement motors in the compact axle are powered by the variable displacement pump and drive the drive wheels via two reduction gears. Braking is naturally obtained by the transmission. (See figure II.4) [13]

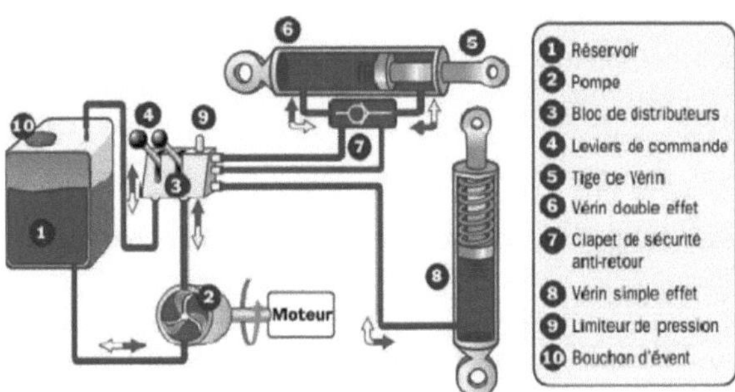

Figure II.4: Hydraulic circuit

II.4.3.3. Management

The steering is hydrostatic and steers the rear wheels via the steering cylinder. Steering can be operated by forcefully turning the steering wheel when the engine is not running.

II.4.3.4. Electrical system

The electrical system is powered by a 12V DC alternator. A 12V battery is used to start the engine.

II.4.3.5. braking system

The hydrostatic drive acts as a service brake. Two reed brakes in the compact axle act as parking brakes. The reed brakes close automatically when the engine is switched off, and brake the truck automatically when it comes to a standstill. The brake pedal also acts as a parking brake. For this reason, the brake pedal must be mechanically locked when the truck is at a standstill.

II.5. Truck safety equipment

In addition to the equipment listed below, the truck must have a :

a. Load backrest: prevents load components from falling onto the driver's platform. The mesh must be designed according to the smallest components of the loads to be transported.

b. Protector: prevents access to moving mechanical parts when these are located in the immediate vicinity of the driver.

c. Driver restraint system: the truck must be fitted with a driver restraint system.restraint system (seat belt) to protect the driver against the risk of it being crushed between the carriage and the ground during an accident. reversal.

d. Horn: of sufficient power

e. Fire extinguisher

f. Braking circuit: used to stop and hold the vehicle at a standstill.

trolley with its maximum authorised load.

g. An ignition key: or any other device preventing the forklift being used by an unauthorised person (see figure II.5)[13].

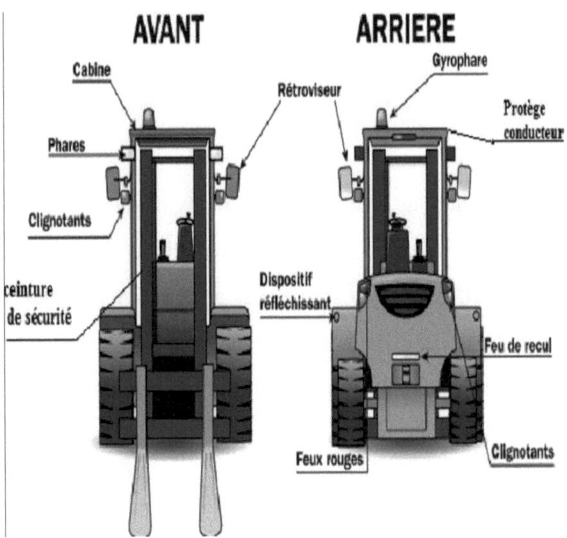

Figure II.5: Forklift truck safety devices.

II.6. Forklift mission

II.6.1. The APTE method

The APTE method (Application aux Techniques d' Entreprise) is an environmental inventory method. It is adapted to the organisation and description of a company's operations and uses a significantly different vocabulary: Service functions (SF) are called basic functions, and include main functions (MF) and constrained functions (CF) [14] :

a. the main functions (PF) represent the purpose of the product's action and are the very expression of need.

b. the constraint functions (CF) reflect the actions and/or reactions of the product in relation to the various external services due to its presence in a system (company) and in the surrounding environment.

c. Technical functions (TF) cover elementary functions and design functions.

II.6.2. Horned beast diagram

The task of the forklift truck differs according to its assignment, which can be: work within the factory, in warehouses, in parts machining workshops[14] .see figure II (6-7-8-9)

Who does it serveWhat does it do?

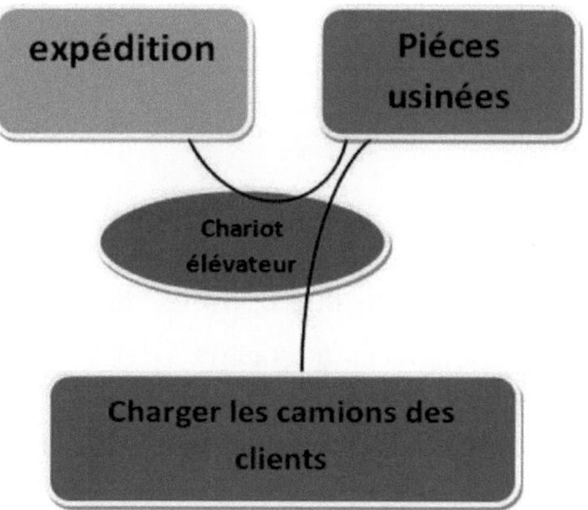

For what purpose?

Figure II.6 Horn-and-horse diagram of the shipping forklift truck

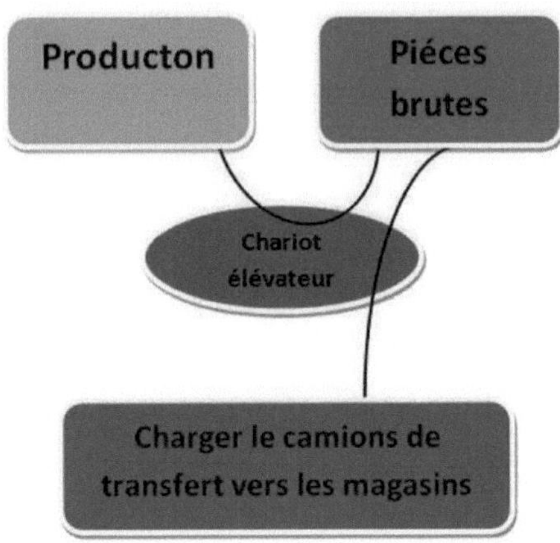

. Figure II.7 Horn-and-beam diagram of the warehouse forklift truck

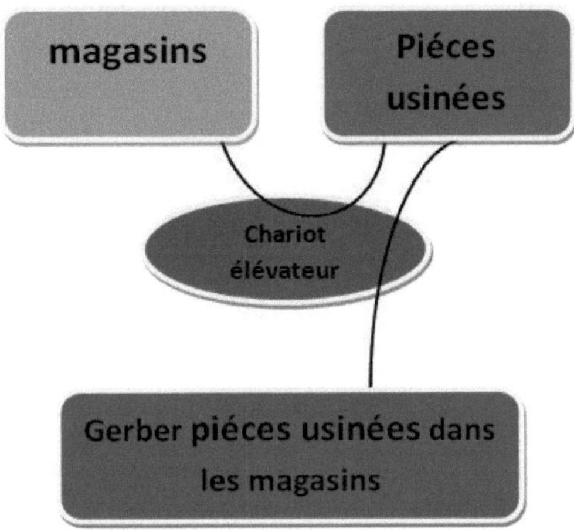

Figure II.8 Horn-and-horse diagram of the factory forklift truck.

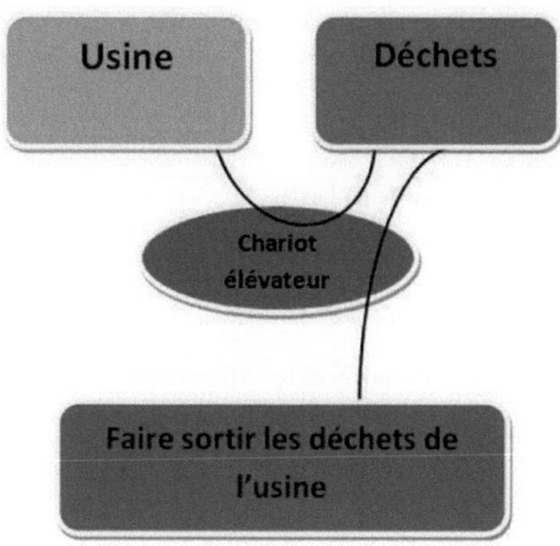

Figure II.9 Horn-and-horse diagram of the waste forklift truck.

II.6.3. Octopus diagram

The octopus diagram is a method of expressing the functions and representing the relationships between the various elements of the surrounding environment and the carriage [14]. (See figure II.10)

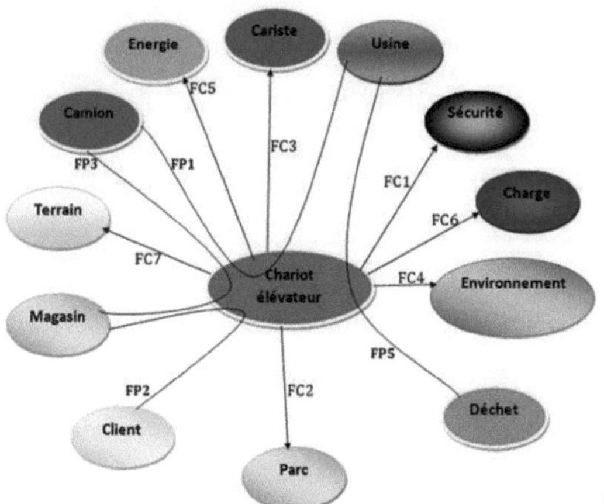

Figure II.10: Octopus diagram of the forklift truck.

The functions of the forklift truck are shown in the following table (Table II.2).

N	The function
FP1	transfer lorries to shops
FP2	Loading customer lorries
FP3	Stacking pallets of ceramic tiles in shops
FP4	Getting waste out of the plant
FC1	Ensuring the safety of forklift drivers and pedestrians
FC2	Comply with maintenance and servicing instructions
FC3	Be easy to handle
FC4	Respecting the environment
FC5	Rationalising fuel consumption
FC6	Stacking and transporting the load
FC7	Ensuring tyres are suited to the ground

Table II.2 Forklift truck functions.

II.7. Regulations and standards

In this section we have set out the results of in-depth research into the regulations, laws and standards applicable to forklift trucks and we have defined the purpose of each regulation and standard.

II.7.1. ASME B56.1 - 1993 Safety Standard for Low Lift and High Lift Trucks

This is an American standard describing the safety rules relating to the training, use, maintenance and design of motorised low-lift and high-lift forklift trucks. For ease of understanding, this standard has been translated into French by the CSST (Norme de sécurité concernant les chariots élévateurs à petite levée et à grande levée, ASME B56.1 1993 A.1995). However, the English edition of this standard is the official version[13].

II.7.2. The Act respecting occupational health and safety (R.S.Q., c. S-2.1)

Its aim is to eliminate at source the dangers and risks that threaten the life, health, safety and physical integrity of workers. It places responsibility for this on workers and employers and gives them the means to participate in achieving this objective through prevention. To this end, it establishes the rights and obligations of workers, employers, owners and suppliers who are subject to it [11].

II.7.3. The Regulation respecting occupational health and safety (R.S.Q., c.S-2.1, r.19.01) [RROHS].

This regulation defines the forklift truck as a lifting device. It prescribes the legal age of use, lays down a compulsory training framework, requires the use of a restraining device for forklift operators and provides a framework for lifting a worker using a forklift[13].

II.7.4. CSA B335 - 04 Safety standard for lift trucks

This Canadian standard is based on ASME B56.1. Application of this standard is voluntary. In addition to prescribing minimum requirements for the training of lift truck operators, this CSA standard sets out specifications for the design, construction, maintenance, inspection and safe operation of lift trucks. The standard also outlines recommended qualifications for trainers of forklift operators as well as technicians and maintenance personnel. [10]

II.8. Identifying the risks associated with using forklift trucks

In the field of occupational risk prevention (table II.3), the International Research and Safety Institute (INRS), which is a scientific and technical body, has set up a database, called EPICEA, designed to record all fatal accidents of significance for prevention[11].

The main accidents occurring during the use of forklift trucks according to the EPICEA database: 226 accidents between 1992 and 2001. (See figure II.11) [15]

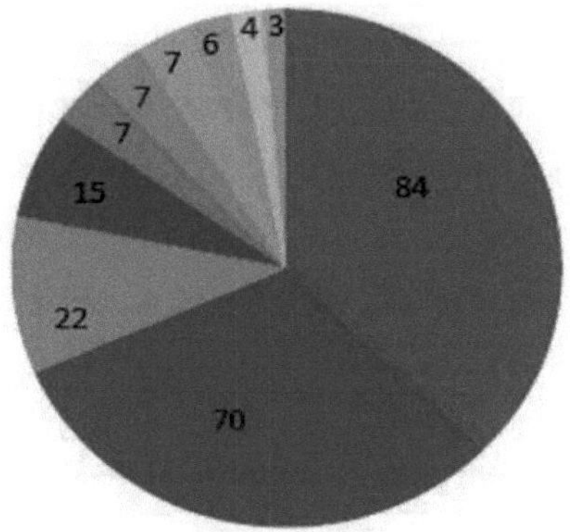

Figure II.11 Percentages of risks associated with the use of forklift trucks

	Pedestrian collision
	Load drop
	Fall from height (person lifted from fork)
	Crushing or jamming of part of the body through the assembly elevator
	Striking of a person by an unmanned forklift truck with engine running
	Front tilting of the truck
	Falling from a platform
	Trolleys falling from ramps or slopes
	Rolling impacts or hitting a wall

Table II.3 Risks associated with the use of forklift trucks

The risk analysis, which takes into account the equipment and its suitability for the installations and the task to be carried out, as well as the workstation environment, has led to the identification of certain risks that have not been the cause of fatal or serious accidents:

a. vibration risk.

b. the risk of harmful emissions from combustion engines.

In **France,** the direct cost of accidents involving forklift trucks is estimated at over €45 million a year, with an annual average of more than 8,000 cases, 500 of which result in permanent disability[15].

In **Algeria**, unfortunately, we do not have any statistics on this type of activity.

types of accident.

CONCLUSION

Our work has enabled us, first of all, to gain a thorough understanding of handling and stacking equipment and their different types and capacities. Secondly, we mastered certain aspects of forklift trucks, including their components, operating principles, the effect of the external environment (risks and accidents associated with the use of this equipment), and the applicable standards for meeting certain conditions.

BIBIOGRAPHICAL

[1] https://www.techniques-ingenieur.fr/base

[2] "le guide de la manutention industriel ", Daniel ISOLA édition Entreprise moderne paris 2003 P 56

[3] Guide to international freight transport, Kamel chaibi

[4] Jean Pierre, CITEAU, Gestion des ressources humaines, 4th edition, Dalloz, Paris, 2002, P189

[5] Claude, PIGANIOL, Techniques et politiques d'améliorations des conditions de travail, Entreprise moderne, Paris, 1980, P XVIII (p48)

[6] Ecole Supérieure des Transports, Graduation thesis, handling: The development of handling, MEMORY presented by ALEXANDRA MERTER, under the direction of Eliane CHACHA, 33rd Class - 2007

[7] the logistics of container port handling terminals [JULIEN DUBREUIL August 2008 CIRRELT

[8] Thesis, advanced handling techniques for the development of container storage problems in a port, Imed ZEMOURI (p 36-52)

[9] Loraine JUNEI " STUDY OF MEANS OF HANDLING ", Academic Master thesis University PARIS EST CRETEIL ,01 JUILET 2017.

[10] INTERNATIONAL STANDARD **ISO 5053-1Second** edition2015-11-01Documentation from the **GERMAN** design office.

[11] The design guide for a forklift truck documentation from the **GERMAN** development office.

[12] **BEDJAMAA ABDELAH** theme study and improvement delarouedentée of an electric forklift c118 Institut National Spécialisé en Formation Professionnelle EL-KHROUB- CONSTANTINE in 2016.

[13] **R'BIGUI Hind** stagethème report**:** Mise en conformité réglementaire et

optimisation technico-économique des chariots élévateursde Super Cérame-Kénitra, Ecole National des Sciences appliqués SIDI MOUHAME BEN ABDELAH MAROQUE Maroc en 2013.

[14] http://methode-apte.fr/les-outils/

[15] http://www.inrs.fr/risques.html.

[16] machining range documentation **GERMAN** sales office.

[17] Method office documentation.

[18] **GUENICHE Ismail** Head of Training Office **GERMAN**.

[19] Article Dimensions of the driving position M.TISSERAND ISTITUT NATIONAL DE RECHERCHE ET DE SECURITE paris.

Printed by Books on Demand GmbH, Norderstedt / Germany